THE

CARPENTER'S AND JOINER'S

HAND-BOOK:

CONTAINING

A COMPLETE TREATISE ON FRAMING HIP AND VALLEY ROOFS.

TOGETHER WITH

MUCH VALUABLE INSTRUCTION FOR ALL MECHANICS AND AMATEURS, USEFUL RULES, TABLES, ETC.,

---

ILLUSTRATED BY THIRTY-SEVEN ENGRAVINGS.

1863.

Copyright © 2018 Read Books Ltd.
This book is copyright and may not be
reproduced or copied in any way without
the express permission of the publisher in writing

British Library Cataloguing-in-Publication Data
A catalogue record for this book is available from
the British Library

Woodworking

Woodworking is the process of making items from wood. Along with stone, mud and animal parts, wood was one of the first materials worked by early humans. There are incredibly early examples of woodwork, evidenced in Mousterian stone tools used by Neanderthal man, which demonstrate our affinity with the wooden medium. In fact, the very development of civilisation is linked to the advancement of increasingly greater degrees of skill in working with these materials.

Examples of Bronze Age wood-carving include tree trunks worked into coffins from northern Germany and Denmark and wooden folding-chairs. The site of Fellbach-Schmieden in Germany has provided fine examples of wooden animal statues from the Iron Age. Woodworking is depicted in many ancient Egyptian drawings, and a considerable amount of ancient Egyptian furniture (such as stools, chairs, tables, beds, chests) has been preserved in tombs. The inner coffins found in the tombs were also made of wood. The metal used by the Egyptians for woodworking tools was originally copper and eventually, after 2000 BC, bronze - as ironworking was unknown until much later. Historically, woodworkers relied upon the woods native to their region, until transportation and trade innovations made more exotic woods available to the craftsman.

Today, often as a contemporary artistic and 'craft' medium, wood is used both in traditional and modern styles; an excellent material for delicate as well as forceful artworks. Wood is used in forms of sculpture, trade, and decoration including chip carving, wood burning, and marquetry, offering a fascination, beauty, and complexity in the grain that often shows even when the medium is painted. It is in some ways easier to shape than harder substances, but an artist or craftsman must develop specific skills to carve it properly. 'Wood carving' is really an entire genre itself, and involves cutting wood generally with a knife in one hand, or a chisel by two hands - or, with one hand on a chisel and one hand on a mallet. The phrase may also refer to the finished product, from individual sculptures to hand-worked mouldings composing part of a tracery.

The making of sculpture in wood has been extremely widely practiced but survives much less well than the other main materials such as stone and bronze, as it is vulnerable to decay, insect damage, and fire. It therefore forms an important hidden element in the arts and crafts history of many cultures. Outdoor wood sculptures do not last long in most parts of the world, so we have little idea how the totem pole tradition developed. Many of the most important sculptures of China and Japan in particular are in wood, and the great majority of African sculptures and that of Oceania also use this medium. There are various forms of carving which can be utilised; 'chip carving' (a style of carving in which knives or chisels are used to remove small chips of the material), 'relief carving' (where figures are carved in a flat panel of wood), 'Scandinavian flat-plane' (where figures are carved in large flat planes, created primarily using a carving knife - and rarely rounded or sanded afterwards) and 'whittling' (simply carving shapes using just a knife). Each of these techniques will need slightly varying tools, but broadly speaking, a specialised 'carving knife' is essential, alongside a 'gouge' (a tool with a curved cutting edge used in a variety of forms and sizes for carving hollows, rounds and sweeping curves), a 'chisel' and a 'coping saw' (a small saw, used to cut off chunks of wood at once).

Wood turning is another common form of woodworking, used to create wooden objects on a lathe. Woodturning differs from most other forms of woodworking in that the wood is moving while a stationary tool is used to cut and shape it. There are two distinct methods of turning wood: 'spindle turning' and 'bowl' or 'faceplate turning'. Their key difference is in the orientation of the wood grain, relative to the axis of the lathe. This variation in orientation changes the tools and techniques used. In spindle turning, the grain runs lengthways along the lathe bed, as if a log was mounted in the lathe. Grain is thus always perpendicular to the direction of rotation under the tool. In bowl turning, the grain runs at right angles to the axis, as if a plank were mounted across the chuck. When a bowl blank rotates, the angle that the grain makes with the cutting tool continually changes between the easy cuts of lengthways and downwards across the grain to two places per rotation where the tool is cutting across the grain and even upwards across it. This varying grain angle limits some of the tools that may be used and requires additional skill in order to cope with it.

The origin of woodturning dates to around 1300 BC when the Egyptians first developed a two-person lathe. One person would turn the wood with a rope while the other used a sharp tool to cut shapes in the wood. The Romans improved the Egyptian design with the addition of a turning bow. Early bow lathes were also developed and used in Germany, France and Britain. In the Middle Ages a pedal replaced hand-operated turning, freeing both the craftsman's hands to hold the woodturning tools. The pedal was usually connected to a pole, often a straight-grained sapling. The system today is called the 'spring pole' lathe. Alternatively, a two-person lathe, called a 'great lathe', allowed a piece to turn continuously (like today's power lathes). A master would cut the wood while an apprentice turned the crank.

As an interesting aside, the term 'bodger' stems from pole lathe turners who used to make chair legs and spindles. A bodger would typically purchase all the trees on a plot of land, set up camp on the plot, and then fell the trees and turn the wood. The spindles and legs that were produced were sold in bulk, for pence per dozen. The bodger's job was considered unfinished because he only made component parts. The term now describes a person who leaves a job unfinished, or does it badly. This could not be more different from perceptions of modern carpentry; a highly skilled trade in which work involves the construction of buildings, ships, timber bridges and concrete framework. The word 'carpenter' is the English rendering of the Old French word *carpentier* (later, *charpentier*) which is derived from the Latin *carpentrius;* '(maker) of a carriage.' Carpenters traditionally worked with natural wood and did the rougher work such as framing, but today many other materials are also used and sometimes the finer trades of cabinet-making and furniture building are considered carpentry.

As is evident from this brief historical and practical overview of woodwork, it is an incredibly varied and exciting genre of arts and crafts; an ancient tradition still relevant in the modern day. Woodworkers range from hobbyists, individuals operating from the home environment, to artisan professionals with specialist workshops, and eventually large-scale factory operations. We hope the reader is inspired by this book to create some woodwork of their own.

PREFACE.

This work has been undertaken by the author to supply a want long felt by the trade: that is, a cheap and convenient "Pocket Guide," containing the most useful and necessary rules for the carpenter.

The writer, in his progress "through the mill," has often felt that such a work as this would have been of great value, and some one principle here demonstrated been worth many times the cost of the book.

It is believed, therefore, that this book will commend itself to those interested, for the reason that it is cheap, that it is plain and easily understood, and that it is useful.

CONTENTS.

	ART.
To find the lengths and bevels of hip and common rafters..................................	1
To find the lengths, &c., of the jacks..............	2
To find the backing of the hip...................	3
Position of the hip-rafter.........................	4
Where to take the length of rafters...............	5
Difference between the hip and valley roof.........	6
Hip and valley combined........................	7
Hip-roof without a deck........................	8
To frame a concave hip-roof.....................	9
An easy way to find the length, &c., of common rafters.......................................	10
Scale to draw roof plans........................	11
To find the form of an angle bracket..............	12
To find the form of the base or covering to a cone...	13
To find the shape of horizontal covering for domes..	14
To divide a line into any number of equal parts.....	15
To find the mitre joint of any angle..............	16
To square a board with compasses................	17
To make a perfect square with compasses..........	18
To find the centre of a circle....................	19
To find the same by another method..............	20
Through any three points not in a line, to draw a circle...	21
Two circles being given, to find a third whose area shall equal the first and second.................	22

CONTENTS.

	ART.
To find the form of a raking crown moulding	23
To lay out an octagon from a square	24
To draw a hexagon from a circle	25
To describe a curve by a set triangle	26
To describe a curve by intersections	27
To describe an elliptical curve by intersection of lines	28
To describe the parabolic curve	29
To find the joints for splayed work	30
Stairs	31
To make the pitch-board	32
To lay out the string	33
To file the fleam-tooth saw	34
To dovetail two pieces of wood on four sides	35
To splice a stick without shortening	36
The difference between large and small files	37
Piling wood on a side-hill	38
To find the number of gallons in a tank	39
To find the area of a circle	40
Capacity of wells and cisterns	41
Weights of various materials	42

THE CARPENTER'S AND JOINER'S HAND-BOOK.

HIP AND VALLEY ROOFS.

The framing of hip and valley roofs, being of a different nature from common square rule framing, seems to be understood by very few. It need scarcely be said, that it is very desirable that this important part of a carpenter's work should be familiar to every one who expects to be rated as a first-class workman. The system here shown is proved, by an experience of several years, to be perfectly correct and practicable; and, as it is simple and easily understood, it is believed to be the best in use. Care has been taken to extend the plates so as to demonstrate each position or principle by it self, so that the inconvenience and confusion of many lines and letters mixed up with each other may be avoided.

ARTICLE 1.—*To find the lengths and bevels of hip and common rafters.*

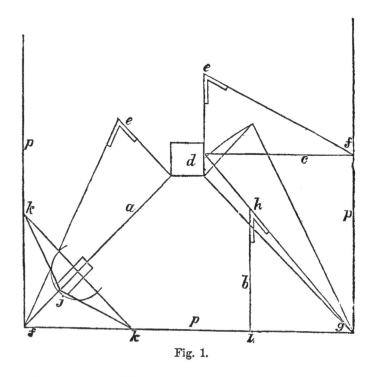

Fig. 1.

Let *p p p* (Fig. 1) represent the face of the plates of the building; *d*, the deck-frame:

is the seat of the hip-rafter; *b*, of the jack; and *c*, of the common rafter. Set the rise of the roof from the ends of the hip and common rafter towards *e e*, square from *a* and *c*; connect *f* and *e*, then the line from *f* to *e* will be the length of the hip and common rafter, and the angles at *e e* will be the down bevels of the same.

2. *To find the length and bevel of the jack-rafters.*

b (Fig. 1) is the seat of a jack-rafter. Set the length of the hip from the corner, *g*, to the line on the face of the deck-frame, and join it to the point at *g*. Extend the jack *b* to meet this line at *h*; then from *i* to *h* will be the length of the jack-rafter, and the angle at *h* will be the *top* bevel of the same.

The length of all the jacks is found in the same way, by extending them to meet the line *h*. The *down* bevel of the jacks is the same as that of the common rafter at *e*.

3. *To find the backing of the hip-rafter.*

At any point on the seat of the hip, *a* (Fig.

1), draw a line at right angles to a, extending to the face of the plates at $k\,k$; upon the points where the lines cross, draw the half circle, just touching the line fe; connect the point at j, where the half circle cuts the line a, with the points $k\,k$; the angle formed at j will be the proper backing of the hip-rafter.

It is not worth while to back the hip-rafter unless the roof is one-quarter pitch or more.

4. It is always desirable to have the hip-rafters on a mitre line, so that the roof will all be the same pitch; but when for some reason this cannot be done, the same rule is employed, but the jacks on each side of the hip are different lengths and bevels.

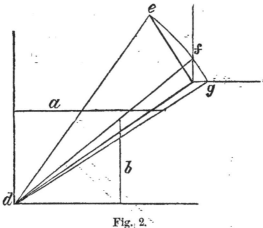

Fig. 2.

The heavy line from *d* (Fig. 2), shows the seat of the hip-rafter; *a* and *b*, the jacks. Set the rise of the roof at *e*; set the length of the hip *d e*, from *d* to *f* on one side of the deck, and from *d* to *g* on the other side; extend the jack *b*, and all the jacks on that side, to the line *d f*, for the length and top bevels; extend the jack *a*, and all on that side, to the line *d g*, for the length and bevels on that side of the hip. The down bevels of the jacks will be the same as that of the common rafters on the same side of the roof.

5. The lengths of hips, jacks, and valley-rafters should be taken on the centre line, and the thickness or half thickness allowed for. (See Fig. 3.)

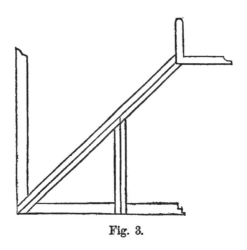

Fig. 3.

6. The valley-roof is the same as the hip-roof inverted. The principle of construction is the same, with a little different application.

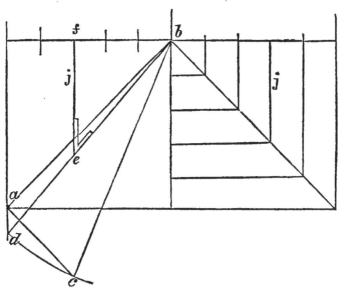

Fig. 4.

Let *a b* (Fig. 4) represent the valley-rafter; *j j* are corresponding jack-rafters. Set the rise of the roof from *a* to *c;* connect *b* and *c:* from *b* to *c* is the length of the valley-rafter, and the angle at *c* the bevel of the same; set the length *b c* on the line from *a;* extend the jack *j* to meet the line *b d* at *e;* then from *e* to *f* is the length of the jack, and the angle at *e* the top bevel of the same.

7. *When the hip and valley are combined,*

so that one end of the jack is on the hip, and the other on the valley.

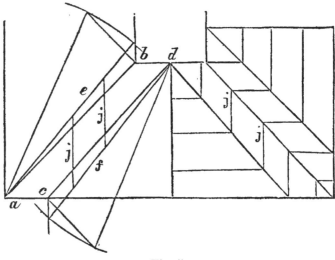

Fig. 5.

a b (Fig. 5) is the hip, and *c d* the valley-rafters. Find the length of each according to the previous directions; find the lines *e* and *f* as before.

Extend the jacks *j j* to the line *e*, for the top bevel on the hip: extend the same on the other end to the line *f*, for the top bevel on the valley; the whole lengths of the jacks is from the line *f* to the line *e*. If the hip and valley rafters lie parallel, the bevel will be the same on each end of the jack.

8. In framing a hip-roof without a decking or observatory, a ridge-pole is used, and of such a length as to bring the hip on a mitre line; but this ridge-pole must be cut half its thickness longer at each end, or the hip will be thrown out of place and the whole job be disarranged.

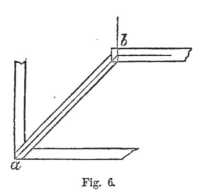

Fig. 6.

This is illustrated by the figure. Suppose the building to be 16 by 20, the ridge would require to be four feet long; but if the stick is four inches thick, for instance, then it

16   THE CARPENTER'S

should be cut four feet four inches long, so that the centre line on the hip, *a*, will point to the centre of the end of the ridge-pole, *b*, at four feet long. This simple fact is often overlooked.

9. *To frame a concave hip-roof.*—(This is much used for verandas, balconies, summer-houses, &c.)

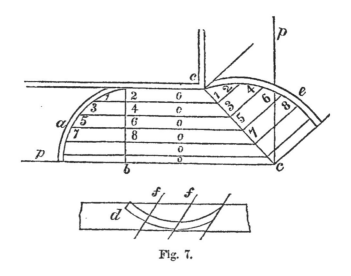

Fig. 7.

To find the curve of the hip.

Let *a* (Fig. 7) be the common rafter in its true position, the line *b* being level. Draw the line *c c*, on the angle the hip-rafter is to lie, generally a mitre line; draw the small lines *o o*, parallel to the plate *p*. The more of these lines, the easier to trace the curve; continue the lines *o o o*, where they strike the line *c c*, square from that line; set the distances 1, 2, 3, 4, &c. (on *a*, from the line *b*) on the line *c c*, towards *e*, at right angles from *c c;* through these points, 2, 4, 6, 8, &c., trace the curve, which will give the form of the hip-rafter.

To get the joints of the jack-rafters, take a piece of plank *d*, (Fig. 7), the thickness required, wide enough to cut a common rafter; mark out the common rafter the full size. Then get the lengths and bevels, the same as a straight raftered roof, which this will be, looking down upon it from above; then lay out your joints from the top edge of the plank, as *f f;* cut these joints first, saw out the curves afterwards, and you will have your jacks all ready to put up. Cut one jack of each length by this method, then use this for a pattern for the others, so as not to waste stuff. It will be seen that the down bevel is different on each jack, *from the curve*, but the same from a straight line, from point to point of a whole rafter.

10. *A quick and easy way to find the lengths and bevels of common rafters.*

Suppose a building is 40 feet wide, and the roof is to rise seven feet. Place your steel square on a board (Fig. 8), twenty inches from the corner one way, and seven inches the other. The angle at *c* will be the bevel of the upper end, and the angle at *d*, the bevel of the lower end of the rafter.

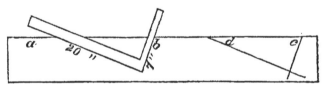

Fig. 8.

11. The length of the rafter will be from *a* to *b*, on the edge of the board. Always buy a square with the inches on one side divided

into twelfths, then you have a convenient scale always at hand for such work as this. The twenty inches shows the twenty feet, half the width of the building; the seven inches, the seven foot rise. Now the distance from *a* to *b*, on the edge of the board, is twenty-one inches, two-twelfths, and one-quarter of a twelfth, therefore this rafter will be 21 feet $2\frac{1}{4}$ inches long.

12. *To find the form of an angle bracket for a cornice.*

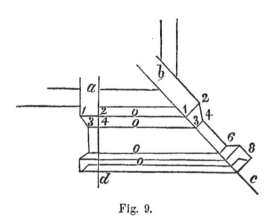

Fig. 9.

Let *a* (Fig. 9) be the common bracket; draw the parallel lines *o o o*, to meet the mitre line *c;* square up on each line at *c*, and set the distances 1, 2, 3, 4, &c., on the common bracket, from the line *d*, on the small lines from *c;* through these points, 2, 4, 6, &c., trace the form of the bracket. This is the same principle illustrated at Fig. 7 and Fig. 20.

13. *To find the form of a base or covering for a cone.*

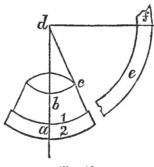

Fig. 10.

Let *a* (Fig. 10) be the width of the base to the cone. Draw the line *b* through the centre of the cone; extend the line of the side *c* till it meets the line *b* at *d;* on *d* for a centre, with 1 and 2 for a radius, describe

e, which will be the shape of the base required; f will be the joint required for the same.

14. *To find the shape of horizontal covering for circular domes.*

The principle is the same as that employed at Fig. 10, supposing the surface of the dome to be composed of many plane surfaces. Therefore, the narrower the pieces are, the more accurately they will fit the dome.

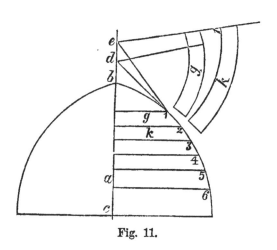

Fig. 11.

Draw the line a through the centre of the dome (Fig. 11); divide the height from b to *c* into as many parts as there are to be courses of boards, or tin. Through 1 and 2 draw a line meeting the centre line at *d*; that point will be the centre for sweeping the edges of the board *g*. Through 2 and 3, draw the line meeting the centre line at *e*; that will be the centre for sweeping the edges of the board *k*, and so on for the other courses.

15. *To divide a line into any number of equal parts.*

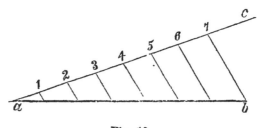

Fig. 12.

Let *a b* (Fig. 12) be the given line. Draw the line *a c*, at any convenient angle, to *a b*; set the dividers any distance, as from 1 to 2. and run off on *a c*, as many points as you wish to divide the line *a b* into; say 7 parts;

connect the point 7 with *b*, and draw the lines at 6, 5, 4, &c., parallel to the line 7 *b*, and the line *a b* will be divided as desired.

16. *To find the mitre joint of any angle.*

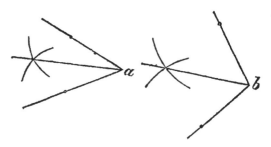

Fig. 13.

Let *a* and *b* (Fig. 13) be the given angles; set off from the points of the angles equals distances each way, and from those points sweep the parts of circles, as shown in the figure. Then a line from the point of the angle through where the circles cross each other, will be the mitre line.

17. *To square a board with compasses.*

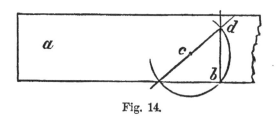

Fig. 14.

Let *a* (Fig. 14) be the board, and *b* the point from which to square. Set the compasses from the point *b* any distance less than the middle of the board, in the direction of *c*. Upon *c* for a centre sweep the circle, as shown. Then draw a straight line from where the circle touches the lower edge of the board, through the centre *c*, cutting the circle at *d*. Then a line from *b* through *d*, will be perfectly square from the lower edge of the board. This is a very useful problem, and will be found valuable for laying out walks and foundations, by using a line or long rod in place of compasses.

18. *To make a perfect square with a pair of compasses.*

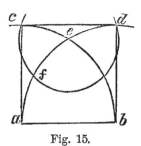

Fig. 15.

Let *a b* (Fig. 15) be the length of a side of the proposed square; upon *a* and *b*, with the whole length for the radius, sweep the parts of circles *a d* and *b c*. Find half the distance from *a* to *e* at *f*; then upon *e* for a centre sweep the circle cutting *f*. Draw the lines from *a* and *b*, through where the circles intersect at *c* and *d*; connect them at the top and it will form a perfect square.

19. *To find the centre of a circle.*

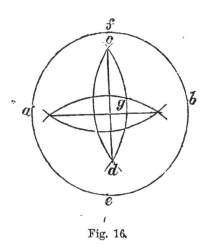

Fig. 16.

Upon two points nearly opposite each other, as *a b* (Fig. 16), draw the two parts of circles, cutting each other at *c d;* repeat the same at the points *e f;* draw the two straight lines intersecting at *g*, which will be the centre required.

20. *Another method.*

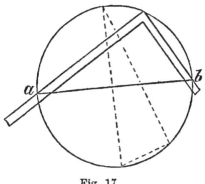

Fig. 17.

Lay a square upon the circle (Fig. 17), with the corner just touching the outer edge of the circle. Draw the line $a\,b$ across the circle where the outside edges of the square touch it. Then half the length of the line $a\,b$ will be the centre required. No matter what is the position of the square, if the corner touches the outside of the circle, the result is the same, as shown by the dotted lines.

21. *Through any three points not in a line, to draw a circle.*

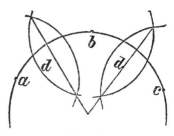

Fig. 18.

Let *a b c* (Fig 18) be the given points. Upon each of these points sweep the parts of circles, cutting each other, as shown in the figure; draw the straight lines *d d*, and where they intersect each other will be the centre required. This method may be employed to find the centre of a circle where but part of the circle is given, as from *a* to *c*.

22. *Two circles being given, to find a third whose surface or area shall equal the first and second.*

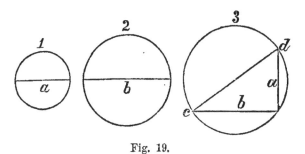

Fig. 19.

Let a and b (Fig. 19) be the given circles. Place the diameter of each at right angles to the other as at 3, connect the ends at c and d, then $c\ d$ will be the diameter of the circle required.

23. *To find the form of a raking crown moulding.*

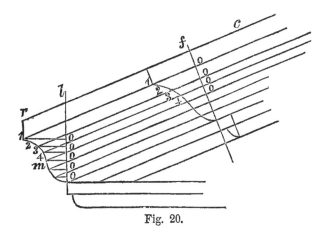

Fig. 20.

m (Fig. 20) is the form of the level crown moulding; *r c* is the pitch of the roof. Draw the line *l*, which shows the thickness of the moulding. Draw the lines *o o o*, parallel to the rake. Where these lines strike the face of the level moulding, draw the horizontal lines 1, 2, 3, &c. Draw the line *f* square from the rake: set the same distances from this line that you find on the level moulding 1, 2, 3, &c. Trace the curve through these points 1, 2, 3, &c., and you have the form of the raking moulding.

Hold the raking moulding in the mitre box, on the same pitch that it is on the roof, the box being level, and cut the mitre in that position.

24. *To make an octagon, or eight-sided figure, from a square.*

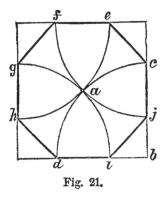

Fig. 21.

Let Fig. 21 be the square; find the centre a; set the compasses from the corner b, to a; describe the circle cutting the outside line at c and d; repeat the same at each corner, and draw lines $c\, e$, $f\, g$, $h\, d$, and $i\, j$. These lines will form the octagon desired.

25. *To draw a hexagon or six-sided figure on a circle.*

Each side of a hexagon drawn within a circle is just half the diameter of that circle. Therefore in describing the hexagon (Fig. 22), first sweep the circle; then without altering the compasses, set off from a to b, from b to c, and so on. Join all these points, a, b, c,

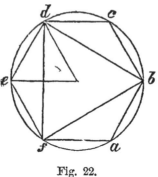

Fig. 22.

&c., and you have an exact hexagon. Join b, d, and f, and you have an equilateral triangle; join d, e, and the centre, and you have another triangle, just one-sixth of the hexagon described.

26. *To describe a curve by a set triangle.*

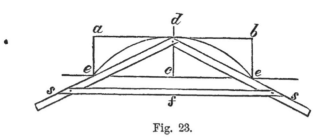

Fig. 23.

Let $a\,b$ (Fig. 23) be the length, and $c\,d$ the height of the curve desired; drive two pins or awls at *e* and *e;* take two strips *s s*, tack them together at *d*, bring the edges out to the pins at *e;* tack on the brace *f*, to keep them in place; hold a pencil at the point *d;* then move the point *d*, towards *e*, both ways, keeping the strips hard against the pins at *e, e*, and the pencil will describe the curve, which is a portion of an exact circle. If the strips are placed at right angles, the curve will be a half circle.

This is a quick and convenient way to get the form of flat centres, for brick arches, window and door heads, &c.

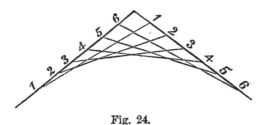

Fig. 24.

27. Fig. 24 shows the method of forming a curve by intersection of lines. If the points 1, 2, 3, &c., are equal on both sides. the curve will be part of a circle.

3*

28. Fig. 25 shows how to form an elliptical curve by intersections. Divide the distance $a\ b$, into as many points as from b to c,

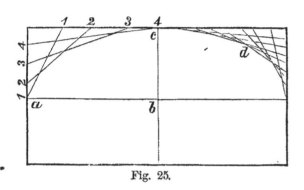

Fig. 25.

and proceed as in Fig. 24. The closer the points 1, 2, 3, &c., are together, the more accurate and clearly defined will be the curve, as at d.

29. Fig. 26 shows the *parabolic* curve.

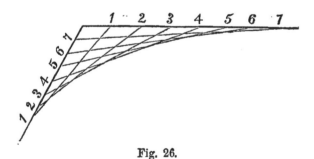

Fig. 26.

This is the form of the curve of the Gothic arch or groin.

30. *To find the joints for splayed work, such as hoppers, trays, &c.*

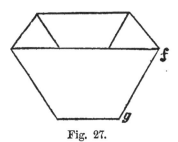
Fig. 27.

Take a separate piece of stuff to find the joints for the hopper, Fig. 27. Strike the bevel $f\,g$, the bevel of the hopper, on the

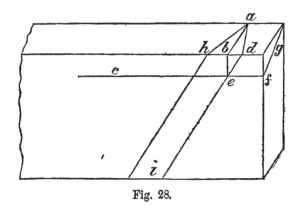
Fig. 28.

end of the piece (Fig. 28); run the gauge-mark c from f; then square on the edge from a, or where you want the outside joint, to b; then square down from b to the gauge-mark c; strike the bevel of the work $f\ g$, from i to d, through the point at e. From a to d will be the joint, the inside corner the longest. If a mitre joint is wanted, set the thickness of the stuff, measuring on $f\ g$, from d to h; the line $a\ h$ will be the mitre joint.

31. *Stairs.**—It is not practicable in a work of this size to go into all the details of stair-building, hand-railing, &c., but a few leading ideas on plain stairs may be introduced.

First, measure the height of the story from the top of one floor to the top of the next; also the run or distance horizontally from the landing to where the first riser is placed.

---

* For a thorough treatise on stair-building in all its details, and many other subjects of interest to the builder, I would recommend "The American House Carpenter," by R. G. Hatfield, New York.

Suppose the height to be 10 ft. 4 in., or 124 inches. As the rise to be easy should not be over 8 inches, divide 124 by 8 to get the number of risers: result, $15\frac{1}{2}$. As it does not come out even, we must make the number of risers 16, and divide it into 124 inches for the width of the risers: result, $7\frac{3}{4}$, the width of the risers. If there is plenty of room for the run, the steps should be made 10 inches wide besides the nosing or projection; but suppose the run to be limited, on account of a door or something else, to 10 ft. 5 in., or 125 inches: divide the distance in inches by the number of steps, which is one less than the number of risers, because the upper floor forms a step for the last riser. Divide 125 by 15, which gives $8\frac{1}{3}$ or $8\frac{4}{12}$ inches, the neat width of the step, which with the nosing, will make about a $9\frac{3}{4}$ step.

32. *To make a pitch-board.*

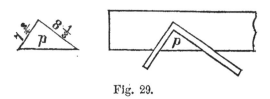

Fig. 29.

Take a piece of thin clear stuff (Fig. 29), and lay the square on the face edge, as shown in the figure, and mark out the pitch-board *p* with a sharp knife.

33. *To lay out the string.*

Nail a piece across the longest edge of the pitch-board, as at *b*, so as to hold it up to the string more conveniently. Then begin at the bottom, sliding the pitch-board along the upper edge of the string, and marking it out, as shown at Fig. 30.

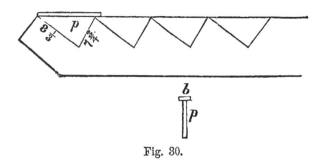

Fig. 30.

The bottom riser must scribe down to the thickness of the step narrower than the others.

34. *To file the fleam-tooth saw.*

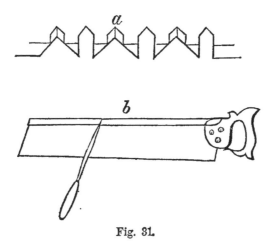

Fig. 31.

Fig. 31 shows the manner of filing the fleam, or lancet toothed saw. *a* shows the form of the teeth, full size; and *b*, the position of holding the saw. The saw is held flat on the bench, and one side is finished before it is turned over. No setting is needed, and the plate should be thin and of the very best quality and temper.

These saws cut at an astonishing rate, cutting equally both ways, and cut as smooth as if the work were finished with the keenest plane.

35. *To dovetail two pieces of wood showing the dovetail on four sides.*

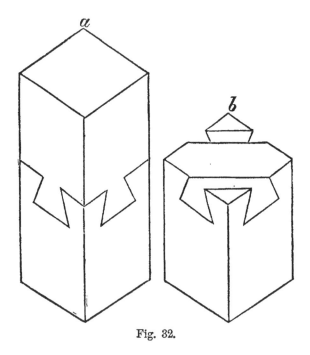

Fig. 32.

a (Fig. 32) shows two blocks joined together with a dovetail on four sides. This looks at first like an impossibility, but *b* shows it to be a very simple matter. This is not of much practical use except as a puzzle. I have seen one of these at a fair attract great attention; nobody could tell how it was done. The two pieces should be of different colored wood and glued together.

36. *To mend or splice a broken stick without making it any shorter or using any new stuff.*

A vessel at sea had the misfortune to break a mast, and there was no timber of any kind to mend it. The carpenter ingeniously overcame the difficulty, without shortening the mast.

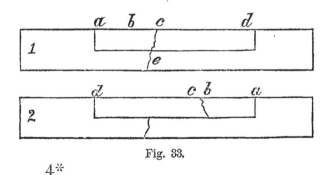

Fig. 33.

e at 1 (Fig. 33) shows where the mast was broken. Cut the piece *a b*, say three feet long, and the piece *c d*, six feet long, half way through the stick. Take out these two pieces, keeping the two broken ends together, turn them end for end, and put them back in place, as shown at 2.

This arrangement not only brought the vessel safe home, but was considered by the owners good for another voyage.

By putting hoops around each joint, the stick would be about as strong as ever.

37. *Is there any difference in the angle of a large or small three-cornered file?*

Certainly not: for the file is an equilateral triangle, equal on all sides.

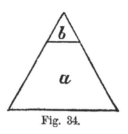

Fig. 34.

Fig. 34 proves this. *a* is a file measuring

one inch on all sides; cut off b, making a file $\frac{1}{4}$ inch on the sides, it will readily be seen that the angle is exactly the same.

Simple as this fact is, it is unknown to many.

38. *Does a pile of wood on a side hill piled perpendicularly, eight feet long, four wide, and four high, contain a cord?*

It does not.

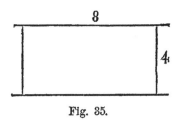

Fig. 35.

To illustrate, let us make a frame (Fig. 35) just 4 by 8 in the clear. When this frame stands level it will hold just a cord.

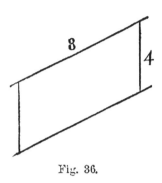

Fig. 36.

Place this frame on a side hill, so as to give it the position in Fig. 36, it will be seen that the 8 ft. sides are brought nearer together, thus lessening its capacity. Continue to increase the steepness of the ground, as at Fig. 37, or more, the 8 ft. sides would finally

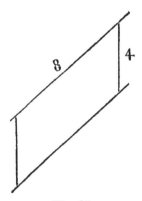

Fig. 37.

come together, and the frame contain nothing at all. It therefore becomes careful buyers of wood to consider where it is piled.

39. *To find the number of gallons in a tank or box*, multiply the number of cubic feet in the tank by $7\frac{3}{4}$.

How many gallons in a tank 8 feet long, 4 feet wide, and 3 feet high?

$$
\begin{array}{r}
8 \\
4 \\
\hline
32 \\
3 \\
\hline
96 \text{ cubic feet.} \\
7\frac{3}{4} \\
\hline
672 \\
72 \\
\hline
\end{array}
$$

Ans. 744 gallons.

40. *To find the area or number of square feet in a circle.*

Three-quarters of the square of the diameter will give the area.

What is the area of a circle 6 ft. in diameter?

$$\begin{array}{r} 6 \\ 6 \\ \hline 36 \\ \frac{3}{4} \\ \hline \end{array}$$

Ans. 27 feet.

For large circles, or where greater accuracy is required, multiply the square of the diameter by the decimal .785.

41. *Capacity of wells and cisterns.*

One foot in depth of a cistern:

3 feet in diameter contains	55¼ gallons.
3½ " " "	75 "
4 " " "	98 "
4½ " " "	124¼ "
5 " " "	153¼ "
5½ feet in diameter contains	185½ "
6 " " "	220¾ "
7 " " "	300½ "
8 " " "	392½ "
9 " " "	497 "
10 " " "	613½ "

A gallon is required by law to contain eight pounds of pure water.

42. *Weights of various materials:*

Material	Lbs. in a cubic foot.
Cast-iron	460
Cast-lead	709
Gold	1,210
Platina	1,345
Steel	488
Pewter	453
Brass	506
Copper	549
Granite	166
Marble	170
Blue stone	160
Pumice-stone	56
Glass	160
Chalk	150
Brick	103
Brickwork laid	95
Clean sand	100
Beech-wood	40
Ash	45
Birch	45
Cedar	28

	Lbs. in a cubic foot.
Hickory	52
Ebony	83
Lignum-vitæ	83
Pine, yellow	38
Cork	15
Pine, white	25
Birch charcoal	34
Pine "	18
Beeswax	60
Water	62½

Lightning Source UK Ltd.
Milton Keynes UK
UKHW040609160120
357068UK00002B/305/P